AF270457

Americana

TELEVISION
THEN AND NOW

Abdo & Daughters
MIDDLE GRADE NONFICTION

An Imprint of Abdo Publishing
abdobooks.com

Jessica Rusick

ABDOBOOKS.COM

Published by Abdo Publishing, a division of ABDO, PO Box 398166, Minneapolis, Minnesota 55439. Copyright © 2024 by Abdo Consulting Group, Inc. International copyrights reserved in all countries. No part of this book may be reproduced in any form without written permission from the publisher. Abdo & Daughters™ is a trademark and logo of Abdo Publishing.

Printed in the United States of America, North Mankato, Minnesota

102023

012024

Design: Kelly Doudna, Mighty Media, Inc.

Production: Denise Hamernik, Mighty Media, Inc.

Editor: Katherine Chu

Cover Photographs: Shutterstock Images (bottom), Wikimedia Commons (top)

Interior Photographs: Alamy Stock Photo, pp. 11, 31, 39; AP Images, pp. 4–5, 18–19, 26–27, 29, 30, 32–33, 36; iStockphoto, pp. 21, 35; Mighty Media, Inc., pp. 3, 42–43; NASA, pp. 6–7; Shutterstock Images, pp. 1 (bottom), 14, 15, 34, 37, 40–41, 44; Wikimedia Commons, pp. 1 (top), 8–9, 12, 13, 16, 17, 20, 22, 23, 24, 25, 28, 45

Design Elements: iStockphoto

LIBRARY OF CONGRESS CONTROL NUMBER: 2023939340

PUBLISHER'S CATALOGING-IN-PUBLICATION DATA

Names: Rusick, Jessica, author.

Title: Television: then and now / by Jessica Rusick

Other title: then and now

Description: Minneapolis, Minnesota : Abdo Publishing, 2024 | Series: Americana | Includes online resources and index.

Identifiers: ISBN 9781098291792 (lib. bdg.) | ISBN 9781098278694 (ebook)

Subjects: LCSH: Americana--Juvenile literature. | Television--Juvenile literature. | Television broadcasting--Juvenile literature. | History, Modern--Juvenile literature.

Classification: DDC 973.0--dc23

TABLE OF CONTENTS

People who don't own
television sets watch the
Apollo 11 rocket launch in a
Sears department store.

ONE GIANT LEAP

It is July 20, 1969. More than 600 million people worldwide are glued to their televisions. In the United States, 93 percent of all households are watching their television screens. Some people crowd around the televisions in their living rooms. In New York City's Central Park, thousands of people gather beneath a large television screen. Everyone is watching one of the biggest events in television history and one of the greatest milestones in human history. Humans are going to land on the moon!

Apollo 11

In 1961, US president John F. Kennedy had set a goal for the United States to be the first country to send humans to walk on the moon. Now, just eight years later, three American astronauts are undertaking the Apollo 11 mission to do just that. The astronauts are Neil Armstrong, Edwin "Buzz" Aldrin, and Michael Collins. Once they land, they will be able to send live footage of their explorations back to Earth with a special video camera.

News anchors in the United States and around the world narrate the moon landing. Many television viewers cheer as the anchors report that the astronauts have successfully landed the landing module, called the *Eagle*, on the moon. Hours later, the world watches the live broadcast of Neil Armstrong taking humankind's first steps onto the moon's surface. "That's one small step for a man, one giant leap for mankind," Armstrong says.

Even though the black-and-white footage is grainy and slightly hard to see, viewers are in awe. A human walking on the moon seemed like a dream only decades ago. But now it is a reality. And television, a

Neil Armstrong working with equipment on the landing module

medium that also seemed impossible only decades ago, has allowed the world to watch.

A Piece of Americana

Television has played a big part in shaping American culture. News broadcasts have influenced how Americans feel about current events, while scripted television shows can either provide an escape from reality or a way to understand it. From its infancy in the 1940s to the streaming era of the twenty-first century, television has both reflected and guided the American public's views. Television is one of the country's most important pieces of Americana.

In 1966, the National Academy of Sciences awarded Vladimir Zworykin the National Medal of Science for his contributions to the instruments of science, engineering, and television.

Chapter 2

TV'S EARLY DAYS

In 1900, Constantin Perskyi, a Russian scientist, invented the term *television*. The word described a theoretical system that transmitted images over telegraph or telephone wires. This was before the development of television technology.

In the 1920s and 1930s, most people listened to news, music, and other entertainment on radios. Several electric companies joined together to form two radio networks, the National Broadcasting Company (NBC) and the Columbia Broadcasting System (CBS). At the same time, inventors had also been developing methods to electronically transmit and receive images over large distances.

One inventor was Russian-born American engineer Vladimir Zworykin. Zworykin worked for the Radio Corporation of America (RCA), the largest electronics and communications company in the United States. RCA realized that television had the potential to become popular and decided to develop the technology.

THE FCC

In 1934, the US Congress established the Federal Communications Commission (FCC). It oversees radio, television, and other types of communication in the United States. Since then, the FCC has regulated the creation of new television networks and what types of content can air on television.

In 1923, Zworykin developed the iconoscope for RCA. The iconoscope was a device that scanned and transmitted images using electrons. In 1929, Zworykin developed the kinescope, a device that could receive images from the iconoscope. But he was not the only one working to invent new television technology. American inventor Philo Farnsworth patented several important television components in the 1920s. RCA wanted to use these components, but Farnsworth refused. After a long court battle, RCA agreed to pay Farnsworth to use his technology in 1939. Because of the legal troubles with Farnsworth, RCA was slow to begin regular public broadcasts.

Broadcast Networks

In the late 1930s and early 1940s, large radio networks expanded into television broadcasting. The networks used large transmitter towers to send television programs over public airwaves. Antennas on top of televisions picked up the signals which were then turned into moving images. The quality of the images depended on how close the television was located to one of the towers. Obstacles like mountains could also negatively affect signals.

One of the leading networks in television broadcasting was the RCA-owned network NBC. It began airing television programs in 1939. NBC's first broadcast was the New York World's Fair opening ceremony, which featured a speech from President Franklin D. Roosevelt. In addition to newsworthy events, the network also aired live sports. At the time, only a few people owned television sets because they were expensive. There also wasn't continual programming, so many people felt televisions were not worth the money. Even though most Americans did not own television sets, they were still in awe of television. Luckily, there were many televisions in

In 1962, people in New York City gathered to watch a speech by US president John F. Kennedy on a television set in a shop's window.

public places, such as store windows and bars. People would gather around storefronts to see the bulky boxes with the small screens or go to bars to watch big sporting events.

World War II

It would take another decade for television to truly catch on in the United States. A big reason for this was World War II, which the United States fought in from 1941 to 1945. During this time, Americans focused on the war effort, so most television set manufacturing declined. Only a few people continued to develop content for television broadcasts during the war.

Once World War II ended, television set manufacturing started again. The RCA Model 630-TS was the first mass-produced TV set. It was sold between 1946 and 1949.

After the war, interest in television increased again. By 1948, it was becoming known as "TV," and there were four major broadcast networks. These were NBC, CBS, the American Broadcasting Company (ABC), and the DuMont Television Network. Initially, networks had a difficult time creating content for this new medium. They often broadcasted live events, such as sports and concerts, because the events were easy and inexpensive to shoot. Short news programs were also common.

Some early television content was also repurposed from radio shows.

DuMont went out of business in the 1950s. CBS, ABC, and NBC were still operating in 2023.

Texaco Star Theater was a radio variety show that aired in the 1930s and 1940s. The live radio program featured comedy and musical performances. The show launched on television in 1948 with American comedian Milton Berle as the host. Berle's funny facial expressions and physical brand of comedy made him perfect for television, and his popularity soon earned him the nickname "Mr. Television."

Texaco Star Theater was television's first hit! The show helped popularize television and as a result, more people began to buy television sets. By 1955, more than half of American homes had one.

Milton Berle was part of a rotating group of hosts on *Texaco Star Theater* when it first aired on TV. He became the permanent host in the fall of 1948.

Sitcoms

As more people purchased TV sets, networks began to produce scripted television shows. One popular form of scripted television was the situation comedy, or sitcom. Sitcoms are 30-minute shows that follow a central cast of characters, such as a group of friends or a family. Before television, sitcoms were popular on the radio. They debuted on television in the late 1940s and early 1950s. Many networks recorded sitcoms in front of live studio audiences, a tradition that is continued by some modern sitcoms.

SIGNING OFF

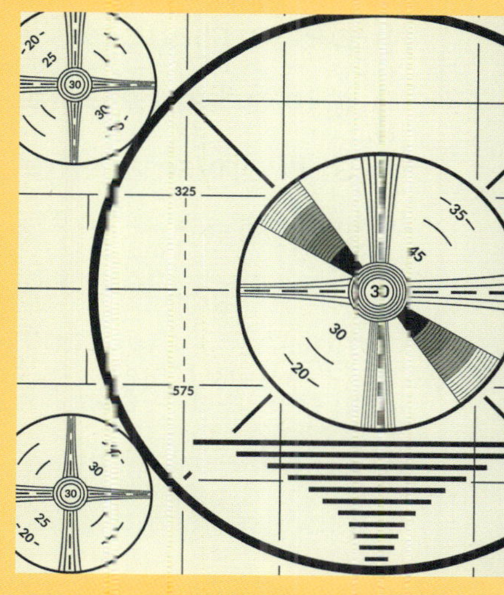

In television's early years, networks did not broadcast at all hours. At night, the networks "signed off" and stopped broadcasting. Television screens then displayed a test pattern until programming started the next day. Even though there were no programs being broadcast, the test pattern indicated that the television was working.

I Love Lucy

The CBS sitcom *I Love Lucy* debuted in 1951. It starred real-life wife and husband Lucille Ball and Desi Arnaz as married couple Lucy and Ricky. The show was a massive hit! For four of its six seasons, it

COLOR TV

Networks filmed and broadcast early television programs in black and white. Although color televisions became available in 1954, they did not become common in American households until the mid-1960s. This is also when networks began filming and broadcasting most shows in color.

was the most popular show on television. Many viewers thought it was both hilarious and heartfelt.

I Love Lucy was also one of the first series that was shot on film instead of being broadcast live. By the end of the 1950s, most television shows were recorded on film, which allowed the networks to replay them on

About 44 million people watched a January 1953 episode of *I Love Lucy*. This included 72 percent of all American households!

air as reruns. Many reruns continued to air for years or even decades after a show had ended, which helped them to gain a wider audience.

Pop Culture Phenomenon

In 1950, only 9 percent of American households had televisions. But during the 1950s, television sets became more widely available and affordable. By 1959, nearly 86 percent of American households had one! In just a few years, television had become an important part of American pop culture, and its influence would only continue to grow.

The 1950s became known as the golden age of television because of television's growth during that period. By 1960, more than 60 million television sets had been sold in the United States.

The first debate between John F. Kennedy (*left*) and Richard Nixon (*right*) was held on September 26, 1960.

Chapter 3

A NEW ERA

By the 1960s, television was no longer just a source of news and entertainment. Television coverage of politics and other newsworthy events began to shape Americans' views. During the 1960 US presidential campaign, several debates between Democrat John F. Kennedy and Republican Richard M. Nixon were broadcast on television. This was the first time presidential debates between major candidates were aired on television.

To many viewers of the first debate, Kennedy appeared confident and handsome while Nixon looked nervous and unkempt. People who watched the debate on television were more likely to say Kennedy won. But people who listened on the radio could not see either candidate's appearance, so they were more likely to say Nixon won. This difference of opinion only proved the power television had to sway public thought. That November, Kennedy won the presidential election after a close race.

Many American news networks broadcast footage of the US war in Vietnam during the 1960s. The actual news reports were often supportive of the US government going to war. But many Americans became disillusioned after seeing the raw, shocking battle footage on television. In 1962, Walter Cronkite became the anchorman for *The CBS Evening News*. Over time, he became known as "the most trusted man in America." In 1968, Cronkite traveled to Vietnam to cover the war. When his commentary aired, it contradicted many positive reports that said the United States could and would be victorious. Cronkite said that, in his opinion, the war would likely end in a stalemate, which led even more Americans to become discontented with the Vietnam War.

On February 20, 1968, Walter Cronkite (*center, with microphone*) interviewed the commanding officer of the First Battalion, First Marines during the Battle of Hue City.

"A Vast Wasteland"

In the 1950s and 1960s, most popular television shows did not reflect what was going on in the real world. *Mister Ed* was a show about a talking horse and his clumsy owner. *Leave It to Beaver* followed the everyday adventures of a boy named Theodore. It often painted an idealized and simplified version of American family life. Westerns that dramatized cowboys in western North America, such as *Gunsmoke* and *Bonanza*, were also popular. Many viewers saw the shows as fun pieces of escapism, but some critics felt the shows lacked intelligence and depth. In 1961, FCC chairman Newton Minnow called television's offerings "a vast wasteland." He was displeased with what he considered to be violent and unbelievable shows.

Westerns became very popular in the 1950s. Between 1949 and 1960, there were more than 100 western series aired on television.

A New Kind of Television

Television content changed over the next decade. In 1969, the most popular show of the year was NBC's sketch comedy *Rowan & Martin's Laugh-In*. It was a fast-paced mix of political humor and zany gags that often addressed real-world issues such as war and racism. CBS noticed *Laugh-In*'s success and updated its television lineup to appeal to a broader audience. This meant releasing three new sitcoms that portrayed complex characters with real-life issues. During the 1970s, CBS released *The Mary Tyler Moore Show*, *All in the Family*, and *M*A*S*H*. Those three series are still considered among the best of all time.

Actor John Wayne (*left*) and musician Tiny Tim (*second from left*) were guest stars on the 100th episode of *Rowan & Martin's Laugh-In*.

The Mary Tyler Moore Show

CBS debuted *The Mary Tyler Moore Show* in 1970. The sitcom was about a single woman, Mary Richards, who worked at a television news station. The show was known for its funny plots and well-written characters. It was also groundbreaking because it explored issues such as sexism in the workplace, divorce, and women's rights.

All in the Family

All in the Family first aired in 1971 and was one of the most successful sitcoms of the 1970s. The comedy followed a working-class family in New York headed by Archie Bunker. Bunker was portrayed as a good person at heart, but he also had many prejudices. His outdated views on race and gender were pitted against those of his more progressive daughter and son-in-law, and were often the butt of the joke. The Bunkers' arguments reflected real conversations that were happening in living rooms across America.

The final episode of *The Mary Tyler Moore Show* aired on March 19, 1977.

M*A*S*H

M*A*S*H premiered in 1972 and featured workers in a US Army Mobile Army Surgical Hospital (MASH). The show was set during the Korean War in the 1950s. But it was often intended as a commentary on the Vietnam War, which the United States was involved in while the show was being made. M*A*S*H was considered a dark comedy because it mixed humor with the hardships of war.

M*A*S*H star Alan Alda (*bottom center*) served in the US Army and was stationed in South Korea in 1956.

Barriers Broken

CBS's shows ushered in a television revolution. Throughout the 1970s, more network shows tackled realistic issues and also started to better represent the American population. For example, most sitcoms in the 1950s featured only white actors. Though some shows such as *Beulah* and *Amos & Andy* starred Black actors, the characters were portrayed in stereotypical ways.

Sitcoms from the 1970s, such as *Sanford and Son*, *Good Times*, and *The Jeffersons*, represented Black characters in more varied and realistic ways. *The Jeffersons* was also the first sitcom to regularly highlight an interracial couple. Although these shows made promising strides, the struggle for equal representation would continue.

The Jeffersons was a spin-off of *All in the Family*. The show received 14 Emmy Award nominations during its ten-year run.

Cable News Network (CNN) was one of the first cable channels.

Chapter 4

THE RISE OF CABLE

The 1980s brought an important shift in television broadcasting. For decades, networks such as NBC, CBS, and ABC had dominated the television landscape. But in the 1980s, the rise of cable TV challenged broadcast's supremacy.

What Is Cable?

Broadcast channels transmit programs over public airwaves which many television sets receive for free. Broadcast channels usually include commercials with their shows and movies. These are advertisements paid for by different companies to promote their products or services. These advertisements help keep the broadcasts free for viewers. Cable television uses cables or satellites to transmit television programs and can transmit more channels than broadcast television. Though most cable TV includes commercials, customers have to pay a monthly fee to watch cable TV.

The Growth of Cable

Though cable TV technology has existed since the 1940s, it was initially just used to send broadcast television to mountainous areas that over-the-air signals couldn't reach. In the beginning, cable TV only had a few original channels. It wasn't very popular because people didn't want to spend money for cable when they could get broadcast television for free. Customers also needed a cable converter box to connect the cable signal to their television set.

But in 1972, the cable channel Home Box Office (HBO) was founded. HBO was a "premium channel" which meant people had to pay extra to receive it. HBO televised recently released movies, did not edit swear words and other objectional material from their programming, and did not include commercials. HBO offered their audience something broadcast television could not, which convinced some people to buy cable.

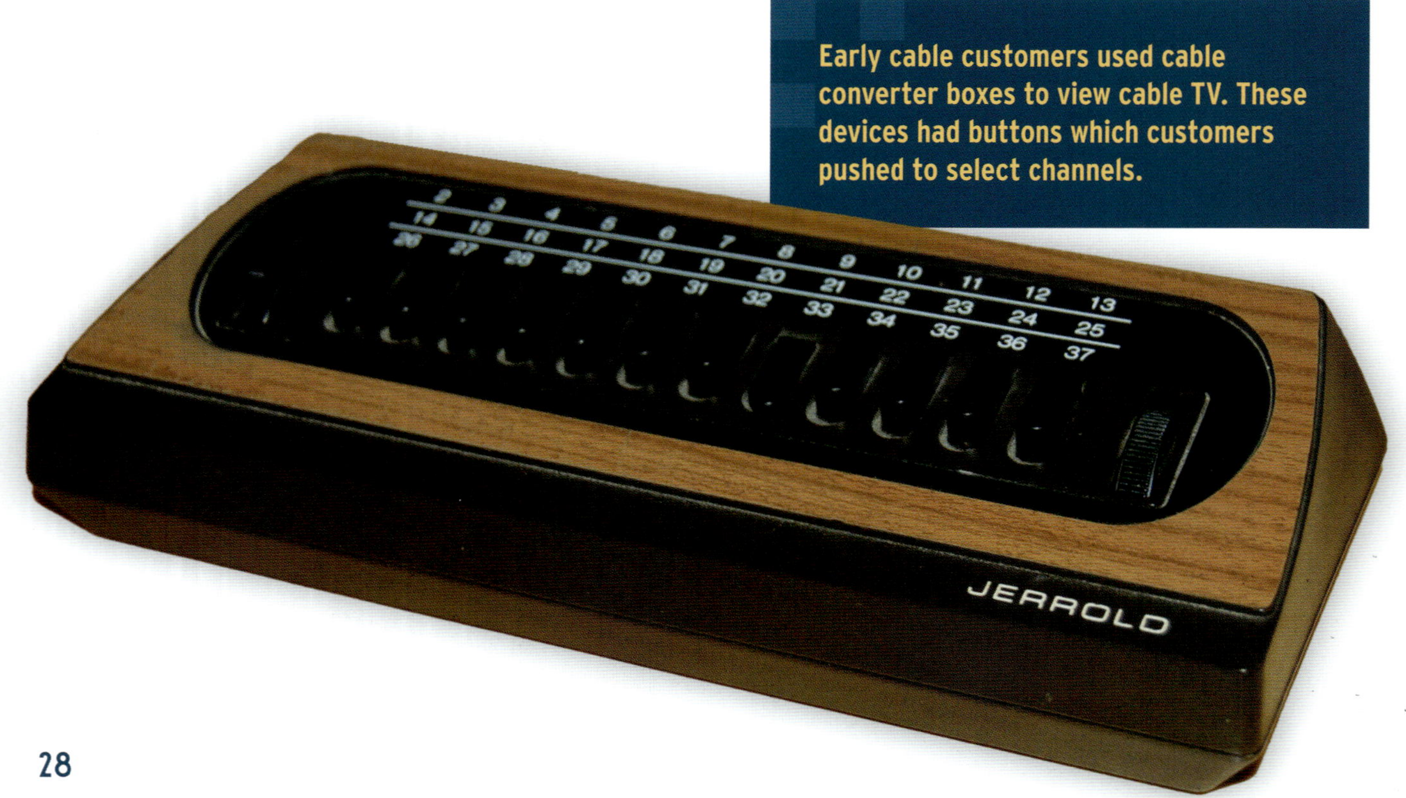

Early cable customers used cable converter boxes to view cable TV. These devices had buttons which customers pushed to select channels.

More cable channels were established in the 1970s and 1980s. These included Nickelodeon and Cable News Network (CNN). Unlike broadcast channels, cable channels catered to specific audiences. CNN offered 24-hour news coverage, while Nickelodeon was exclusively for children's shows.

Over time, cable companies added more channels dedicated to cooking shows, history programs, old movies, and more. In 1980, 23 percent of Americans had cable TV and by 1990, the number had grown to 60 percent. In 1989, 67 percent of Americans watched broadcast television in the evening, which was known as prime time. This number declined over the next decade as cable TV viewership increased.

Nick Jonas gets slimed at the Nickelodeon Kids' Choice Awards in 2015.

Changing Habits

Cable changed how Americans watched television. On average, Americans went from getting a few television channels to more than 50 after installing cable. In the past, tens of millions of people may have watched the same sitcom episode when it aired. But with so many channels and shows now available, Americans had more variety to choose from.

Drama Gets Big

In the 1980s, broadcast networks released high-quality drama shows to attract cable viewers. NBC debuted the cop drama *Hill Street Blues* in 1981. The series had complex characters and well-written storylines. It looked and sounded more like a movie than a television show. Although *Hill Street Blues* was not a popular, mainstream hit, it was a hit with television critics. During its seven-year run, the show

In 1981, the cable channel MTV debuted to show music videos. One of the first videos was for a song by rock singer and songwriter Pat Benatar.

received 98 Emmy Award nominations and 26 wins. *Hill Street Blues* inspired other television networks to air high-quality dramas.

By 1994, dramas had become a popular and mainstream te evision genre. In the late 1990s and early 2000s, cable networks began to debut their own dramas. HBO's *The Sopranos*, considered one of the best and most popular television dramas of all time, debuted in 1999. HBO's *Game of Thrones* and AMC's *The Walking Dead* and *Breaking Bad* were several cable dramas ranked among the best and most well-liked shows of the 2010s.

Hill Street Blues was different from other cop shows. It often touched on race and class issues instead of just being a cop drama.

Netflix DVDs were mailed in distinctive red envelopes.

THE STREAMING ERA

The early 2000s brought in a new era of television. People weren't tied to their televisions. They could watch television anywhere on their computers, phones, and other devices. Just like cable TV had in the 1980s, internet streaming was about to take over American television.

Netflix Leads the Way

Netflix was founded in 1997 as a DVD rental company. For a monthly fee, customers could rent television shows and movies on DVD and have them shipped to their homes. In 2007, Netflix introduced online streaming. In addition to renting DVDs, customers could also watch television shows and movies instantly over the internet. By 2010, more Netflix customers streamed video instead of renting the DVDs. Other popular streaming services included Amazon Unbox, which debuted in 2006, and Hulu in 2007. Amazon Prime Video replaced Amazon Unbox in 2011.

In 2013, Netflix began to release original content. Experts in the streaming industry saw this as a risky and expensive move. But when Netflix's first major original TV show, *House of Cards*, debuted in 2013, it was a groundbreaking success! The first season earned nine Emmy nominations. *House of Cards* was also the first streaming show to receive such recognition from the Television Academy. The show's popularity proved that streaming services could compete with broadcast and cable TV networks when it came to making acclaimed television shows.

House of Cards was a political drama set in Washington, DC. Michel Gill (*left*) played the US president and Jayne Atkinson (*right*) played a US senator.

More Original Content

Netflix continued to make original shows after *House of Cards*. In 2022, its most successful series included *Stranger Things*, *Bridgerton*, and *Squid Game*. Netflix also paved the way for other streaming services to make original content. These included Hulu's series *The Handmaid's Tale* and Amazon's series *The Kicks*. By 2022, Emmy nominations for streaming content were more common, and streaming content was a major part of pop culture.

As of 2023, Netflix had more than 230 million subscribers, the most of any streaming service.

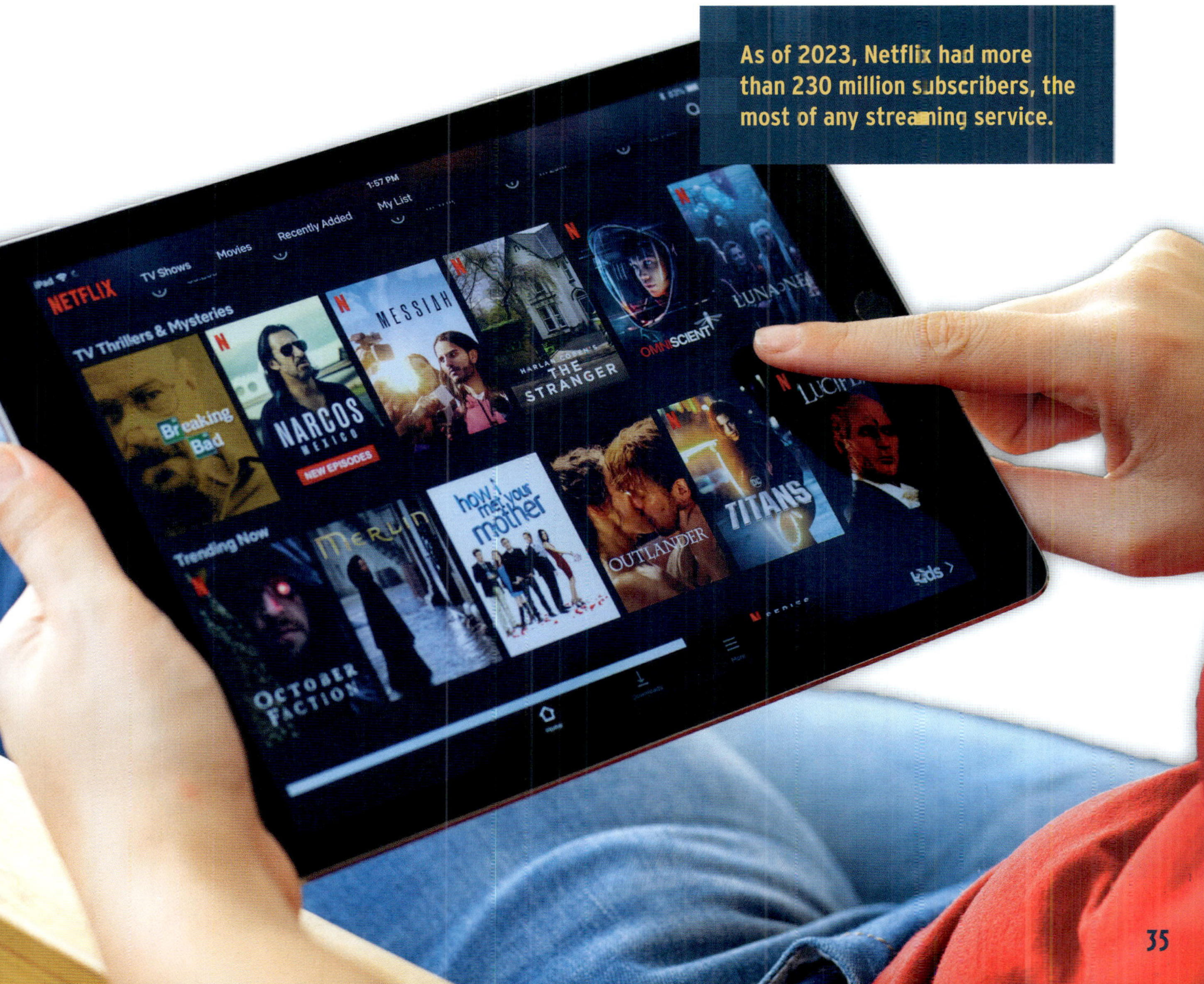

Streaming Changes the Game

Streaming services changed the way Americans watched television in several ways. Many streaming services would release an entire season of a television show at once. With broadcast or cable TV, people had to watch each episode weekly when it aired. With streaming services, people could watch a whole show in just a few sittings, which became known as binge-watching.

Streaming services also allowed television to become portable. In addition to watching on their televisions, people could stream television on their computers, phones, and other devices. Traditional forms of television have become less popular because streaming services provide a more convenient alternative. For example, rather than watching shows when they aired on network TV, people could stream the shows on Hulu the next day.

In addition, some streaming services aired live television, such as news and sporting events, while being less expensive than cable packages. In 2022, for the first time ever, more viewers watched television on streaming services than on broadcast or cable TV.

Many Netflix programs and stars have won Emmy Awards, including Claire Foy, who appeared in the Netflix drama *The Crown.*

New Streaming Services

More companies released streaming services in the late 2010s and early 2020s, taking advantage of their exclusive content to attract customers. In 2019, the Walt Disney Company debuted its streaming service Disney+. At that time, the company owned the rights to all content made by Disney Studios, Marvel Studios, and Lucasfilm Ltd., which produces the Star Wars franchise. Disney+ also created original television shows based on these properties. These included the Marvel series *WandaVision* and the Star Wars series *The Mandalorian*. In its first year, Disney+ gained more than 70 million subscribers

When HBO's streaming service HBO Max debuted in 2020, it featured a large collection of its movies and original television shows. To stay competitive, some television networks stopped licensing popular shows to outside streaming services and created their own streaming platforms. NBC's *The Office* was one of Netflix's most-watched shows. When Netflix's streaming rights to the show expired in 2020, NBC moved *The Office* to its new streaming service, Peacock, in January 2021. The move was a success! After *The Office* premiered on Peacock, the service saw a 5 percent increase in subscribers.

Diversity on Television

In recent decades, television has grown better at reflecting the makeup of the US population. By 2021, a wide variety of shows across broadcast, cable, and streaming platforms represented more diversity. Shows featuring diverse casts and writers generally became more popular. There has been a slow increase in the number of women, disabled people, and LGBTQ+ community members included in new shows. Shows also represented Black actors at a rate consistent with the proportion of Black Americans. But many shows still underrepresented Latino, Asian American, and Native American actors. Television writers and show staff also had few people of color.

What's Next?

By 2022, there were more than 200 streaming services in the United States. Each had different licensed and original content to offer. Just as people paid for cable packages with many channels, many people now pay for multiple streaming services. In the future, it is likely that streaming services will continue to become more successful than broadcast and cable TV. It is also likely that streaming companies will buy other streaming platforms to stay competitive.

New technology may also change the way Americans watch television. Instead of watching on a screen, future viewers may use virtual reality (VR) headsets or goggles. This technology will allow viewers to feel like they are in their favorite television shows!

In January 2021, the Disney+ series *WandaVision* was the most viewed show across all streaming video services. It stars Elizabeth Olsen as the Marvel superhero Scarlet Witch.

Television Then and Now

Television has gone through many changes in its short history and will continue to develop in the future. But no matter where or how Americans watch television, it is sure to remain an integral part of American pop culture. New television shows will continue to delight viewers, and reruns and streaming services will ensure that older television shows find new fan bases.

VR is still being developed as a way to show movies and TV shows. One day, watching TV could include all five of the viewer's senses!

What can you create that is inspired by television? Start by thinking about your favorite channels, shows, characters, and themes. Then put your imagination to work!

Television and snacks go perfectly together. **Turn dollar store caddies into handy TV trays**. Decorate them with duct tape. Then line them with coffee filters and fill them with your favorite TV snacks.

Turn a cardboard box into **a television that can hold your art supplies**. Cover it with duct tape and add hooks for hanging scissors. Glue cardboard tubes to the bottom for legs.

Make a mini retro TV to use as a phone stand. Cover a small box with duct tape. Glue on beads for knobs and wires for the antennas.

TIMELINE

1920s
Vladimir Zworykin develops the iconoscope and kinescope for RCA. These devices are key to electronic television systems.

1948
Texaco Star Theater airs for the first time. It becomes television's first hit.

1959
Nearly 86 percent of American households have a television.

1939
NBC begins airing regular public television broadcasts.

1951
The first episode of the sitcom *I Love Lucy* airs.

1969
About 600 million people watch the Apollo 11 moon landing on television.

1970-1972

CBS debuts *The Mary Tyler Moore Show, All in the Family*, and *M*A*S*H*. These shows are among the first to tackle real-life issues.

1997

Netflix is founded as a DVD rental and delivery service.

1972

HBO is founded.

1980s

Cable TV grows in popularity.

2007

Netflix starts offering a streaming service.

2013

Netflix produces its first major original show, *House of Cards*.

2022

More viewers watch television on streaming services than on broadcast or cable TV.

GLOSSARY

academy—an organization specializing in knowledge of a particular subject.

acclaimed—having received approval or praise.

component—one of the parts or units of a combination, mixture, or system.

consistent—unchanging over time.

contradict—to say the opposite of a statement.

debate—a public discussion about a question or topic.

debut—to make a first appearance.

disillusioned—having lost faith or trust in something formerly regarded as good or valuable.

diversity—having or including people of different races or cultures.

escapism—a form of entertainment that provides an escape from reality or everyday matters.

franchise—a series of related works, such as TV shows, movies, or video games, that feature the same characters.

genre—a type of art, music, or literature.

hilarious—very funny.

idealized—representing persons or things as one believes they should be rather than as they are.

integral—needed in order to be complete.

Korean War—a war fought in North and South Korea from 1950 to 1953. The US government sent troops to help South Korea.

portray—to present an image of something or someone.

scripted—following a script. A script is the written words and directions used to put on a play, movie, or television show.

stalemate—a drawn or undecided contest.

stereotypical—based on a widely held but overly simple idea about a group.

subscriber—someone who signs up to receive something on a regular basis.

Vietnam War—from 1954 to 1975. A long, failed attempt by the United States to stop North Vietnam from taking over South Vietnam.

Booklinks
NONFICTION NETWORK
FREE! ONLINE NONFICTION RESOURCES

To learn more about television, please visit **abdobooklinks.com** or scan this QR code. These links are routinely monitored and updated to provide the most current information available.

INDEX